SpringerBriefs in Applied Sciences and Technology

Continuum Mechanics

W0079457

Series Editors

Holm Altenbach, Institut für Mechanik, Lehrstuhl für Technische Mechanik, Otto von Guericke University Magdeburg, Magdeburg, Sachsen-Anhalt, Germany

Andreas Öchsner, Faculty of Mechanical Engineering, Esslingen University of Applied Sciences, Esslingen am Neckar, Germany

These SpringerBriefs publish concise summaries of cutting-edge research and practical applications on any subject of Continuum Mechanics and Generalized Continua, including the theory of elasticity, heat conduction, thermodynamics, electromagnetic continua, as well as applied mathematics.

SpringerBriefs in Continuum Mechanics are devoted to the publication of fundamentals and applications, presenting concise summaries of cutting-edge research and practical applications across a wide spectrum of fields. Featuring compact volumes of 50 to 125 pages, the series covers a range of content from professional to academic.

Robert J. Reidel · Paul Reidel · Hamid Garmestani ·
Omar S. Es-Said

On the Abnormal/Coarse Grain Formation in K-Monel 500 Alloy

 Springer

Robert J. Reidel
The Boeing Company
Huntington Beach, CA, USA

Hamid Garmestani
School of Materials Science
and Engineering
Georgia Institute of Technology
Atlanta, GA, USA

Paul Reidel
Photometrics, Inc.
Huntington Beach, CA, USA

Omar S. Es-Said
Mechanical Engineering Department
Loyola Marymount University
Los Angeles, CA, USA

ISSN 2191-530X   ISSN 2191-5318 (electronic)
SpringerBriefs in Applied Sciences and Technology
ISSN 2625-1329   ISSN 2625-1337 (electronic)
SpringerBriefs in Continuum Mechanics
ISBN 978-3-031-31078-2   ISBN 978-3-031-31079-9 (eBook)
https://doi.org/10.1007/978-3-031-31079-9

This Springer imprint is published by the registered company Springer Nature Switzerland AG
The registered company address is: Gewerbestrasse 11, 6330 Cham, Switzerland

Contents

List of Figures

List of Tables

Chapter 1
Introduction

Abnormal grain growth (AGG) or secondary recrystallization or discontinuous grain growth or exaggerated grain growth is a coarsening type of microstructure where some large grains grow unusually quickly in a matrix of fine grains [1, 2]. The onset of AGG depends on crystallographic texture, surface effects, grain boundary faceting/de-faceting and second phase particles [1–4]. The weakening of the pinning effects of second phase particles (solute atoms and trapping pores) results in the inhomogeneous grain growth [5–7].

AGG may occur because the driving force is the reduction in grain or sub grain boundary energy of the solid, [8–9]. It also may occur because of the favorable crystallographic orientation of certain grains, for example, Goss orientation {110} < 001 > in Fe-3% Si steel [10, 11]. Another example is strong cube texture attributed to the initial AGG of {100} oriented grains in strip cast Fe Si electrical steel [12]. However, AGG was also observed in Al-3.5% Cu with a weak crystallographic texture [13].

A study [1] was conducted to analyze the AGG of 304 stainless steel. It was observed that at low temperatures normal grain growth occurred, at higher temperature (850 and 900 °C) a transition to AGG occurred. This was attributed to carbide particles coarsening. After 1 or 2 h annealing, normal grain growth occurred. After 3 or 4 h, growth transitioned to AGG. After long anneal times (12 h) the growth returned to normal. At very high annealing temperatures ($0.8\ T_{mp}$), AGG again was observed. This was attributed to grain boundary faceting/defaceting. The use of ultrafine grained (UFG) Ni, which exhibits super plasticity was meeting difficulties in keeping the ultrafine grains without AGG occurring.

It was concluded that AGG was a result of nonlinear migration of faceted boundaries with respect to the driving force [14].

The objective of this research project was to study the factors that possibly generate abnormal grain growth (AGG) during the processing of K-Monel 500 nickel alloy. During processing, AGG always has the potential to occur, which can be detrimental to the mechanical properties of the material and on occasion results in failed products

R. J. Reidel et al., *On the Abnormal/Coarse Grain Formation in K-Monel 500 Alloy*, SpringerBriefs in Continuum Mechanics, https://doi.org/10.1007/978-3-031-31079-9_1

issues [15]. The factors to be studied are very fast heating rate solution treatments and the effects of different quench rates on the age hardening behavior of hot-rolled and cold-drawn initial microstructures. The very large grains referred to by industry personnel as "germinated" or "elephant" grains are referred to as coarse grains in this paper.

Monel K-500 is a precipitation hardenable nickel-copper–aluminum alloy, and was a registered trademark of INCO Alloys International, Inc. This alloy was known for possessing excellent corrosion resistance, coupled with greater strength and hardness over other comparable alloys. The significance of the aluminum and titanium alloying elements was manifested when Ni_3 (Al, Ti) precipitated out into the nickel matrix under specified heat treatments. This precipitated, intermetallic phase is commonly referred to as gamma prime and has a face-centered cubic structure similar to that of the matrix and a lattice constant having 1% or less mismatch with the matrix. Typical applications for this alloy are pump shafts and impellers, oil-well drill collars and instruments, electronic components, gyros, fasteners, springs, and surgical blades [16, 17].

Chapter 2
Experimental

2.1 Materials

Two bars of Monel alloy K-500 were received from INCO Alloys International located in Huntington, West Virginia, U.S.A., and their chemical composition is shown in Table 2.1 [16]. The specified composition identifies the chemical composition required per the Federal Specification (QQ-N-00286F(SH)) governing this alloy. The actual composition, provided by INCO Alloys, identifies the chemical composition of the melt of Monel alloy K-500 from which the samples used in this study were obtained. These comprised a 50.8 cm (20″) long piece of 1.9 cm (0.75″) diameter hot-rolled stock and a 91.4 cm (36″) long piece of 1.60 cm (0.63″) diameter cold-drawn stock. The cold-drawn stock was drawn from the hot-rolled stock with 30% cold work.

The as-received materials were cut into 3.8 ± 0.08 cm ($1.5″ \pm 1/32″$) long sections for both the hot-rolled and cold-drawn materials. In addition, a 0.16 cm (1/16″) diameter hole was drilled at the center of each sample to the center-line axis of the part as illustrated in Fig. 2.1. The mass of the hot-rolled and cold-drawn samples, on average, were 93.7 gms (0.21 lbs) and 64.0 gms (0.14 lbs), respectively. The sample designations and processing sequence for both the hot-rolled and cold-drawn materials are shown in Fig. 2.2.

2.2 Heat Treatments

The cold-drawn and hot-rolled samples 1 and 2 (Fig. 2.2) were placed in separate lots in a vacuum furnace for solution heat treatment. The treatment comprised of ramping the temperature from ambient to 1010 ± 14 °C (1850 ± 25 °F) and holding for 30 min, while a thermocouple, that was used to monitor the sample temperature, was inserted in the 0.16 cm (1/16″) diameter hole of sample 1. The vacuum furnace was maintained

© The Author(s), under exclusive license to Springer Nature Switzerland AG 2023
R. J. Reidel et al., *On the Abnormal/Coarse Grain Formation in K-Monel 500 Alloy*, SpringerBriefs in Continuum Mechanics,
https://doi.org/10.1007/978-3-031-31079-9_2

Table 2.1 Chemical composition, % by weight, of Monel Alloy K-500

Chemical composition	Specified [16]	Actual
Constituent	(%)	(%)
Nickel (plus Cobalt), minimum	63.0	65.33
Copper	27.0–33.0	29.65
Aluminum	2.30–0.85	3.89
Titanium	0.35–0.85	0.46
Iron, maximum	2.0	0.73
Manganese, maximum	1.5	0.72
Silicon maximum	0.5	0.08
Carbon, maximum	0.18	0.14
Sulfur, maximum	0.01	< 0.001

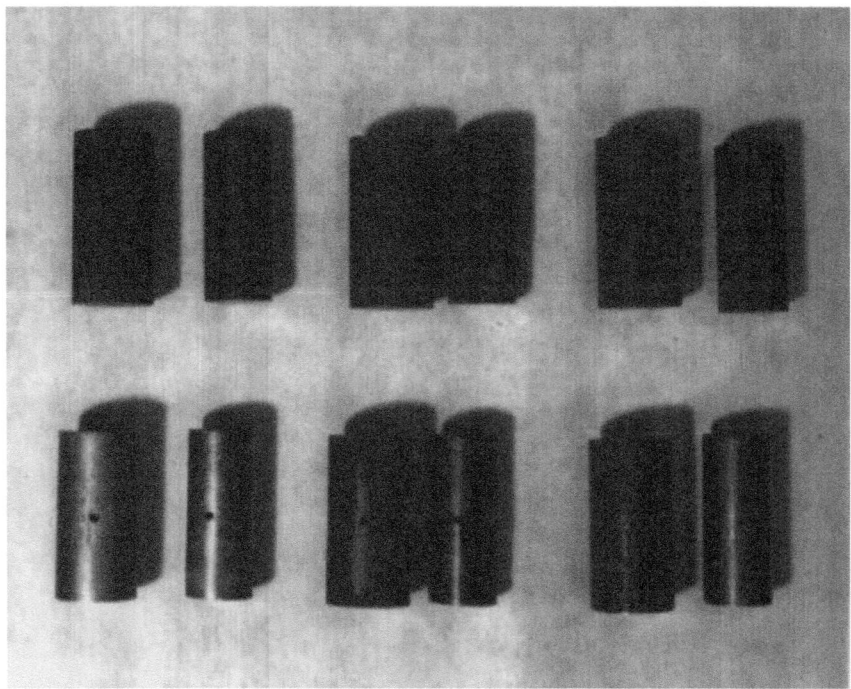

Fig. 2.1 Illustration of hot-rolled and cold-drawn test samples

at a pressure of 266.6 N/m^2 (2×10^{-5} torr) for the duration of the solution treat and cooled-down to approximately 315 °C (600 °F), after which an argon gas was back-filled into the chamber quickly, to bring the temperature to ambient. These two sets of samples were solution treated in the vacuum furnace for two reasons. First, the furnace provided a very slow means of "quenching" the samples, and second, the

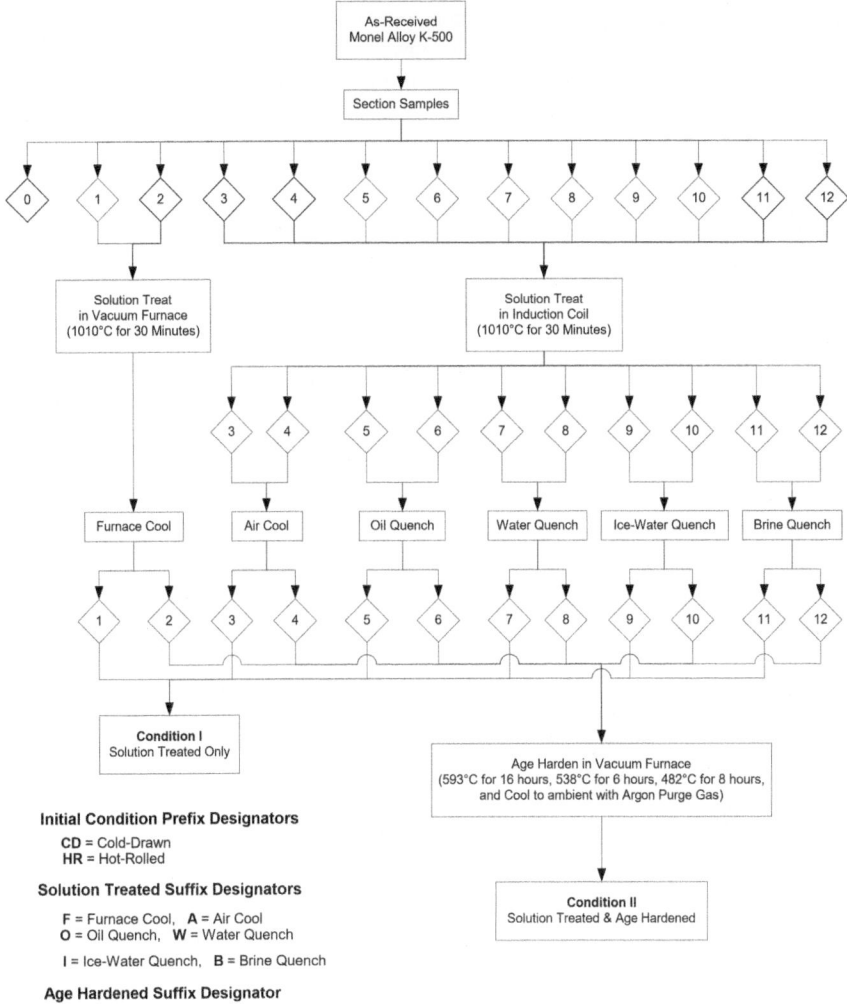

Fig. 2.2 Heat treatment flow chart and sample identification

vacuum furnace method closely resembled a typical manufacturing heat treatment method.

Each of the remaining samples (3 through 12, Fig. 2.2) for both the hot-rolled and cold drawn materials were solution treated at $1010 \pm 14\,°C$ ($1850 \pm 25\,°F$) for 30 min by induction heat treating. The induction coil that the samples were suspended on was comprised of 0.64 cm (1/4″) diameter copper tubing with a 5.24 cm (2 1/16″)

outer diameter, 4.29 cm (1 11/16″) inner diameter and was 9.19 cm (2 7/16″) high. A 5.3 KV DC setting on the Cycle-Dyne power supply was equated to a temperature of 1010 °C (1851 °F), the induction coil frequency was 450 KHz, the power input supplied was 5.2 DC kV and 5.3 DC kV for the cold-drawn and hot-rolled samples, respectively, and the current was 0.82 DC amperes. This equated to a power density of 2247.5 kW/m^2 (1.45 kW/in^2) and 1906.50 kW/m^2 (1.23 kW/in^2) for the cold-drawn and hot-rolled samples, respectively. The induction coil test set-up is provided in Fig. 2.3a and b. Induction heat treating was selected as the solution treatment method for all the remainder of the samples (Fig. 2.2) since it facilitated a test set-up that permitted an instantaneous quench of the samples following the conclusion of the solution treatment temperature dwell.

The quench medium, whether it was oil, water, ice water or brine, was positioned directly below the induction coil, and each sample was suspended within the induction coil by a shielded thermocouple that was inserted into the 0.16 cm (1/16″) diameter hole in the side of the sample (Fig. 2.3a). Data acquisition of the thermal measurements was obtained utilizing a computer, a type "K" thermocouple, and an Omega type "K" cold junction compensator. The thermal data was collected at two-minute intervals during the dwells, and nine times every second during the quench. Upon the completion of the 30-min dwell at the solution treatment temperature, the samples were immediately immersed in the appropriate quench medium (Fig. 2.3b). An illustration of typical cooling curves for the various quench media is provided in Fig. 2.4a, and a blow-up of the first 4-min of the quench is illustrated in Fig. 2.4b.

The oil quench medium was 24 °C (75 °F) Castrol R-100 cutting oil, and the ice-water quench medium was comprised of ice mixed with water 15 min prior to use and used when the temperature of the bath uniformly measured 0 °C (32 °F). The 22 °C (72 °F) brine quench medium was comprised of a 20% by weight salt-water solution and was vigorously stirred just prior to the immersion of the sample into the solution to make the solution homogeneous prior to quenching. The solution treatment specification used for all samples was Federal Specification QQ-N-00286F for Nickel-Copper-Aluminum Alloy, Wrought [16].

The even numbered samples (2, 4, 6, 8, 10 and 12, Fig. 2.2) for both the hot-rolled and cold-drawn samples were placed in a vacuum furnace for age hardening. Correspondingly, each odd numbered sample preceding an even number designated a solution treated only sample subjected to the same quench medium. The age hardening parameters were [16]: (a) heat to 593 ± 14 °C (1100 ± 25 °F) and hold for 16 h, (b) furnace cool to 538 ± 14 °C (1000 ± 25 °F) at 14 °C/h (25 °F/h) and hold for 6 h, (c) furnace cool to 482 ± 14 °C (900 ± 25 °F) at 14 °C/h (25 °F/h) and hold for 8 h, and (d) back-fill the vacuum furnace with argon gas to bring the chamber to ambient. Argon was used in lieu of air in a vacuum furnace to preclude oxidation of the internal elements (and parts) that would occur if air were used. The age hardening thermal profile is illustrated in Fig. 2.4c.

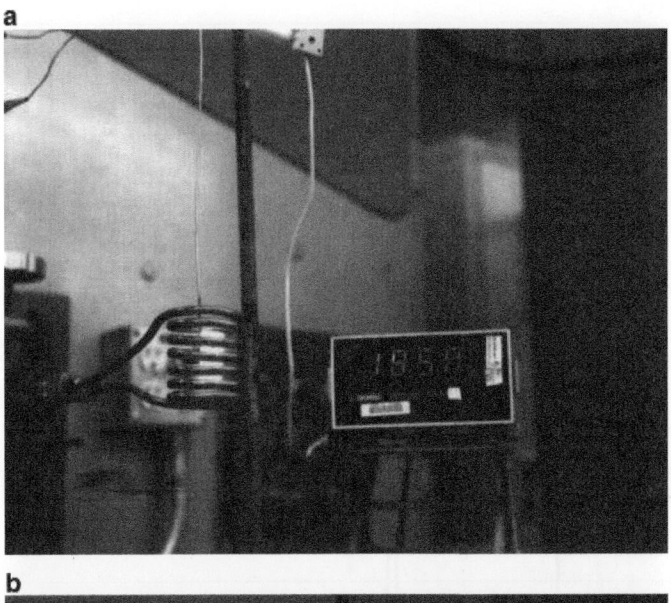

Fig. 2.3 **a** Close-up view of sample in induction coil. **b** Close-up view of sample released into quench medium

2.3 Characterization and Testing

Longitudinal and transverse samples to the rolling direction were for microhardness testing and were mounted in epoxy and used for scanning electron microscopy after the hardness measurements were taken. The samples used for transmission electron

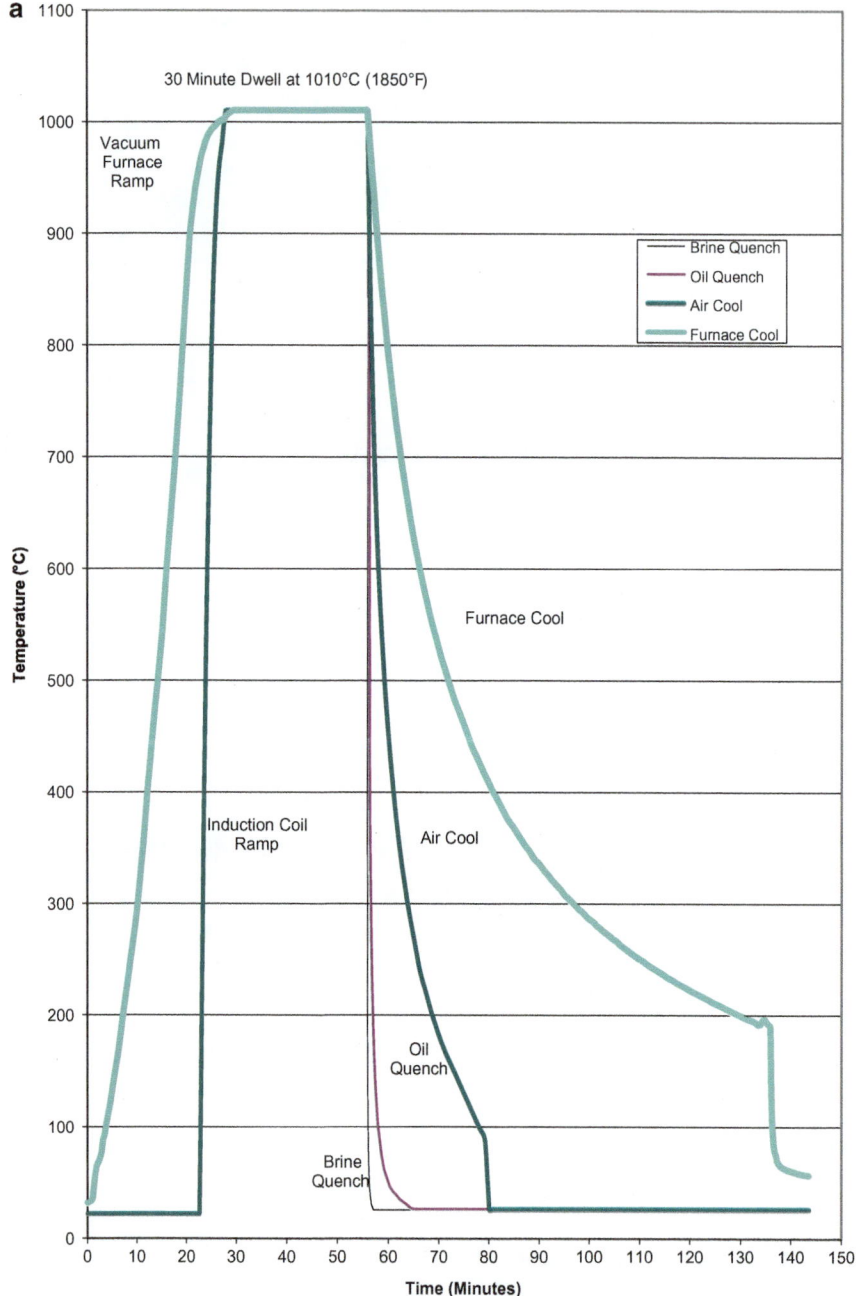

Fig. 2.4 **a** Solution treat cooling curves. **b** Blow-up of solution treat cooling curves. **c** Age hardening thermal profile

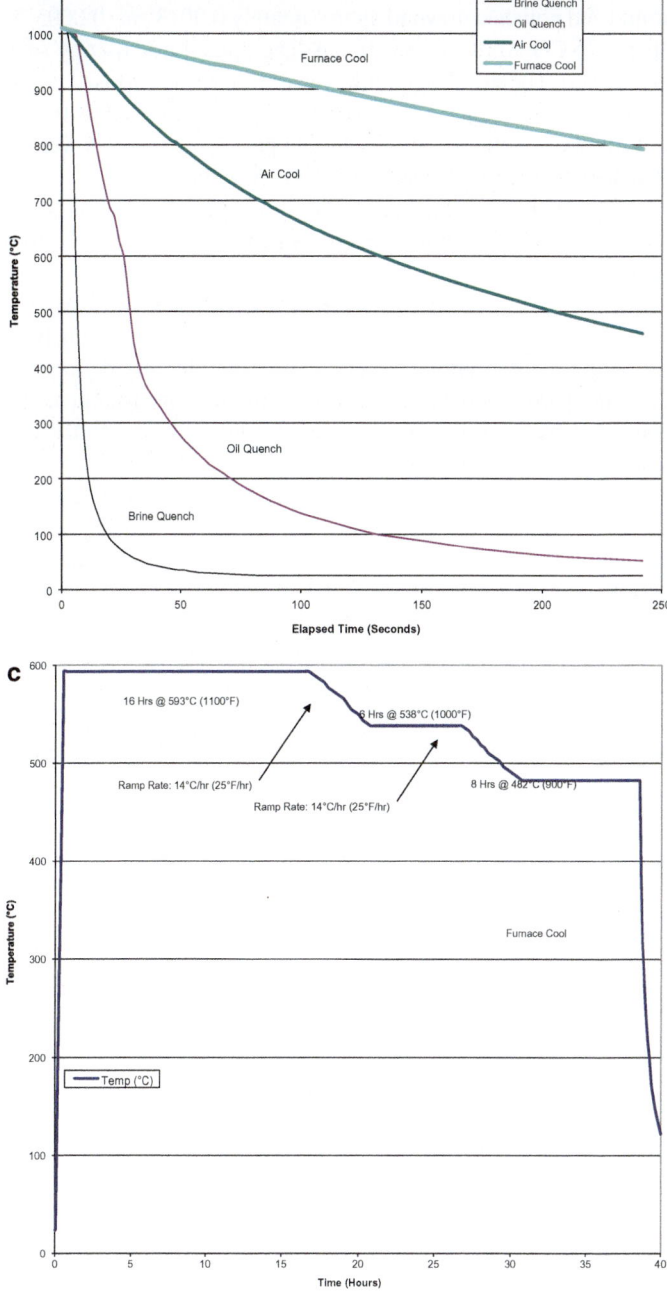

Fig. 2.4   (continued)

microscopy were cut to be 0.08 cm (0.03″) thick by electro discharge machining (EDM), ground and polished down to approximately 0.0013 cm (0.0005″) thick, and then supplied to INCO Alloys International. The optical microscopy samples were ground and polished. The metallographic preparation was comprised of grinding and polishing phases. The sequence of sandpaper grits used was 120, 240, 320, 400 and finally 600. With the conclusion of the sanding phase, the samples were thoroughly rinsed, blown dry and were coarse polished using 6-μ diamond paste. Metallographic etching was performed immediately following the fine polishing phase, the etchant being a 10% solution of ammonium persulfate mixed at a ratio of one-to-one with a 10% solution of sodium cyanide.

The microhardness measurements were taken on a Shimadzu Micro Hardness Tester model HMV-2000, with the parameters for the measurements being: Knoop scale, 500 g load, and a 15-s application time. The samples tested were fine polished prior to taking the hardness measurements, the hardness measurements were taken on a Rockwell Hardness Tester, and the samples tested were sanded to a 600-grit finish prior to taking the hardness measurements.

Grain sizing was performed on all samples from 400× optical micrographs, and only on the "fine" (versus coarse) grained regions of the samples. The grain sizes were obtained for both the transverse and longitudinal axes using the Line Intercept Method according to the equation:

$$DD = (LLTT)/(PP * MM)$$

where D is the grain size in microns (μm), L_T is the total test line length, P is the number of intercepts, and M is the magnification of the micrograph [18]. Optical macrographs were typically taken at 3.5×, 4.25×, 25× and 50×, and optical micrographs were typically taken at 100×, 400× and 1000×.

The samples were analyzed using an Amray model 1910FE field emission scanning electron microscope. SEM micrographs were typically taken at 400×, 1000× , 2500×, 60,000× and 150,000×, while the micrographs at 400× and 1000× were taken for a direct comparison with the optical micrographs, and the 60,000× and 150,000× micrographs were taken for their illustration of the gamma-prime precipitate in the aged samples, or the lack thereof in the as-received or solution-treated only samples. The samples were also analyzed using a Transmission Electron Microscope (TEM), whose micrographs were typically taken at 8000×, 22,000×, 28,000×, 48,000× and 60,000×. The TEM work was performed by INCO Alloys International, Huntington, West Virginia.

To determine the crystallographic texture and preferred orientation distribution functions (ODF), selected samples were studied by using a Scintag 2-C-X1 Advanced Diffraction system and POPLA software [19].

Chapter 3
Results and Discussion

3.1 Quenching Rates

To quantify the effect of the varying quench media on the cooling rates of the test samples, thermal data was collected on the samples at a rate of 9 data points per second during the cooling process from the solution treat temperature to well below temperatures at which age hardening occurs. The quench rates for the test samples in ascending order were approximately: furnace cool (0.6 °C/s), air cool (3 °C/s), oil quench (20 °C/s), water quench (47 °C/s), ice-water quench (92 °C/s) and brine quench (171 °C/s). The samples were maintained as near as possible to the target temperature of 1010 °C (1850 °F) during the solution treatment [16]. Upon conclusion of the 30-min solution treat dwell, the samples were immersed in the liquid medium (oil, water, ice-water, or brine). The fluid in immediate proximity to the sample was turned to a gas during the very violent heat transfer. This gas layer provided a temporary insulating layer between the hot sample and the cold fluid, and within a few seconds, the gas layer was minimized, the heat was readily transferred to the quench fluid, and the part was cooled.

3.2 Mechanical Properties

Both the Knoop microhardness and Rockwell hardness measurements resulted in the same trends with regards to the hardness properties of the material. Both the solution treatments of the cold-drawn and hot-rolled materials removed the work/strain imposed on the microstructure. This was readily apparent with approximately 47% and 17% decreases in hardness for the as-received cold-drawn and hot-rolled materials, respectively. The reduced microhardness data for the various sample conditions is provided in Table 3.1, while the Rockwell hardness data for the various sample conditions is provided in Table 3.2. Regardless of the starting condition of the material (cold-drawn or hot-rolled) the samples were solution treated (Condition I) and then age hardened (Condition II) to essentially the same hardness values for each condition. However, the furnace-cooled samples obtained a small but distinguishable increase in hardness for the solution treated only condition, over the samples quenched by any other faster means. The justification for this occurring is that a small percentage of the gamma-prime precipitated out of solution during the extremely slow quench from the solution treatment temperature, prior to the age hardening heat treatment process. The effects of this partial precipitation were not visibly apparent by scanning electron microscopy. This slight increase in hardness was not maintained after the samples were age hardened. Consequently, a fast or slow cooling rate does not have any specific effect on the softness of annealed solid solution nickel materials, although rapid quenching is preferred for age hardening alloys to ensure maximum precipitation response [20]. The hardness values of the solution treated, and age hardened samples were the same as the values of the 30% cold-worked as-received material.

The age hardening method of attaining the high hardness can be preferred over cold working for both fabricability and stability. In some cases, it is desirable to machine the material in the solution treated condition (softer), and then age harden the material. This approach decreases tool wear and machining time, and when this approach is taken, the shrinkage factor of 2.5×10^{-4} cm/cm, because of aging, must be taken into account for close tolerance parts. If the material is fabricated into a part that is under continuous load, that part is much more stable over an extended period

Table 3.1 Microhardness data for cold-drawn and hot-rolled Monel alloy K-500

Knoop scale	Cold-drawn Monel alloy K-500 as-received: **301**						
500 g, 15 s load	Cooling method	Furnace cool	Air cool	Oil quench	Water quench	Ice-water quench	Brine quench
Material condition							
Solution treated	172	156	159	164	160	166	
Solution treated and aged	304	322	315	329	328	314	
Knoop scale	Hot-rolled Monel alloy K-500 as-received: **181**						
500 g, 15 s load	Cooling method	Furnace cool	Air cool	Oil quench	Water quench	Ice-water quench	Brine quench
Material Condition							
Solution treated	178	143	146	161	149	157	
Solution treated and aged	319	308	315	324	323	323	

Table 3.2 Rockwell hardness data for cold-drawn and hot-rolled Monel alloy K-500

Rockwell "B" scale	As-received hot-rolled Monel alloy K-500: **85** (21C)						
100 Kg, 1/16′ ball	Cooling method	Furnace cool	Air cool	Oil quench	Water quench	Ice-water quench	Brine quench
Material condition							
Cold-drawn solution treated	82 (17C)	78 (17C)	78 (18C)	78 (17C)	79 (17C)	78 (18C)
Hot-rolled solution treated	84 (17C)	73 (14C)	72 (12C)	71 (14C)	74 (14C)	72 (15C)
Rockwell "C" scale	As-received cold-drawn Monel alloy K-500: **28**						
150 Kg, Brale	Cooling method	Furnace cool	Air cool	Oil quench	Water quench	Ice-water quench	Brine quench
Material condition							
Cold-drawn solution treated and aged	32	31	31	33	31	32
Hot-rolled solution treated and aged	31	28	30	31	31	31

Note "C" scale hardness measurements in parentheses provided as reference for comparison

with the hardness obtained by precipitates pinning the dislocations rather than by increasing the number of dislocations by plastic deformation [21].

3.3 Microstructure

The grain size data was obtained from 400× magnification optical micrographs, with the regions of grains evaluated from "fine" versus "coarse" grained, (abnormal grain growth, AGG) regions of the sample. The coarse grains were orders of magnitudes larger in size than the fine or "normal" sized grains, with the grains illustrated in the microstructures of the cold-drawn material, and the grains of similar size in the hot-rolled material being examples of "normal" sized grains. Regardless of the original starting condition of the material or quench rate, no significant variation in grain size was observed for samples in either the transverse or longitudinal directions. The grains were found to be approximately 17 μm in size, is equiaxed, and whose data is provided in Table 3.3.

The cold-drawn material solution treated in an induction coil and both the cold-drawn and hot-rolled material solution treated in a vacuum furnace incurred a small percentage of coarse grains. The quantity of coarse grains incurred for these two latter conditions was less than 5%. Figures 3.1 and 3.2 illustrate the homogeneous

Table 3.3 Grain size data for transverse and longitudinal axes

Grain size (μm) Material condition	Cold-drawn Monel alloy K-500 as-received Trans.: **21** Long.: **18**						
	Cooling method	Furnace cool	Air cool	Oil quench	Water quench	Ice-water quench	Brine quench
Transverse: solution treated	21	20	19	19	15	18
Longitudinal: solution treated	21	19	23	20	17	20
Transverse: solution treated and aged	17	18	17	18	16	17
Longitudinal: solution treated and aged	17	18	18	18	19	17
Grain size (μm) Material condition	Hot-rolled Monel alloy K-500 as-received Trans.: **14** Long.: **16**						
	Cooling method	Furnace cool	Air cool	Oil quench	Water quench	Ice-water quench	Brine quench
Transverse: solution treated	14	18	18	19	16	20
Longitudinal: solution treated	13	14	16	18	18	17
Transverse: solution treated and aged	11	15	14	13	14	15
Longitudinal: solution treated and aged	14	16	14	14	15	15

starting condition of the hot rolled as-received Monel alloy K-500, while Figs. 3.3 and 3.4 illustrate the few islands of coarse grains present in the hot-rolled vacuum furnace solution treated material, and Figs. 3.5, 3.6, 3.7, 3.8, 3.9, 3.10, 3.11, 3.12, 3.13, 3.14, 3.15, through 3.16 illustrate the massive dual microstructure incurred in the hot-rolled Monel alloy K-500 samples solution treated in an induction coil. As a solution treatment check of the hot-rolled as-received material, a sample was placed in a standard air furnace for the 30-min solution treatment and water quenched. In this sample a very homogeneous microstructure was obtained, as illustrated in Figs. 3.17 and 3.18. This sample was outside of the initial scope of the heat treat matrix and was assigned a sample identification of HRWF. The Knoop microhardness and Rockwell hardness of this sample were 174 Knoop and 81B respectively. These hardness values fall within the confines of the values obtained by vacuum furnace and induction coil solution treated samples. This indicates that there might be a strong correlation between the coarse grains (AGG) and the induction coil heat treatment. Typical transverse and longitudinal cross-sections for both the induction coil solution treated hot-rolled and cold-drawn samples are provided in Figs. 3.19 and 3.20, respectively. The severity of the coarse grains is clearly illustrated in the hot-rolled sample (Fig. 3.19), whereas the few coarse grains (AGG) in the cold-drawn sample are barely discernible (Fig. 3.20). This suggest that there is a strong correlation between the coarse grains and the starting hot rolled condition.

Fig. 3.1 Transverse cross-section optical macrograph of hot rolled as-received Monel alloy K-500. Sample HR0. Etch: NaCN, $(NH_4)_2S_2O_8$ (4.25×)

Fig. 3.2 Transverse cross-section optical micrograph of hot-rolled as-received Monel alloy K-500. Sample HR0. Etch: NaCN, $(NH_4)_2S_2O_8$ (400×)

Fig. 3.3 Transverse
cross-section optical
macrograph of hot-rolled,
vacuum furnace solution
treated and furnace cooled
Monel alloy K-500. Sample
HR1F. Etch: NaCN,
$(NH_4)_2S_2O_8$ (4.25×)

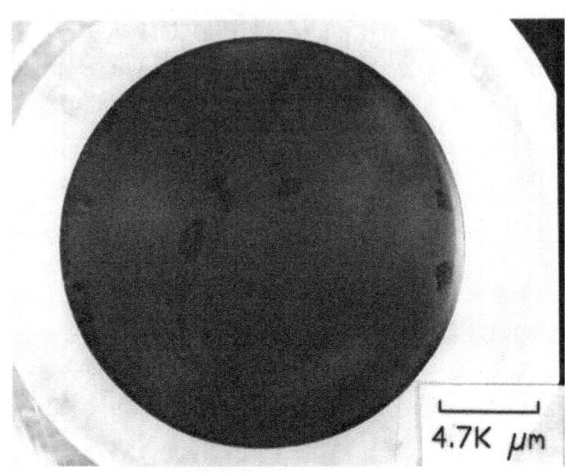

Fig. 3.4 Transverse cross-section optical micrograph of hot-rolled, vacuum furnace solution treated and furnace cooled Monel alloy K-500. Sample HR1F. Etch: NaCN, $(NH_4)_2S_2O_8$ (50×)

A macrograph for the as-received cold-drawn material is provided in Fig. 3.21, and micrographs illustrating the homogeneous transverse and longitudinal axes are provided in Figs. 3.22 and 3.23, respectively. The grains are essentially equiaxed, with a very faint sign of the direction of cold-working present in the longitudinal cross-section. Figures 3.24 and 3.25 illustrate the difference in microstructure presented with varying etches. The NaCN, $(NH_4)_2S_2O_8$ etch clearly reveals the annealing

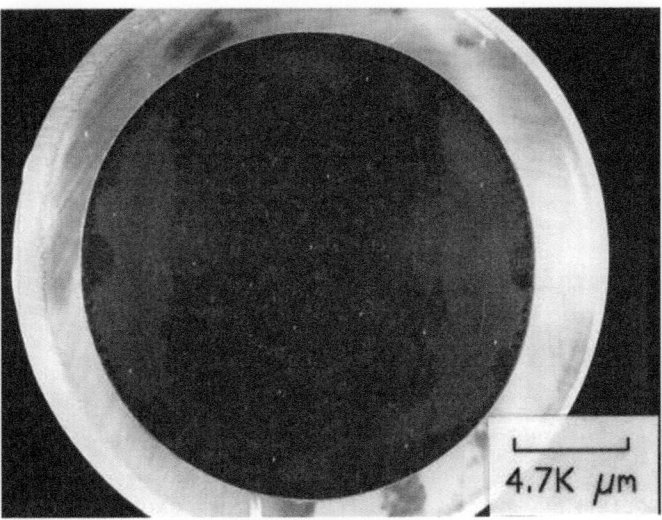

Fig. 3.5 Transverse cross-section optical macrograph of hot-rolled, induction coil solution treated and air-cooled Monel alloy K-500. Sample HR3A. Etch: NaCN, $(NH_4)_2S_2O_8$ (4.25×)

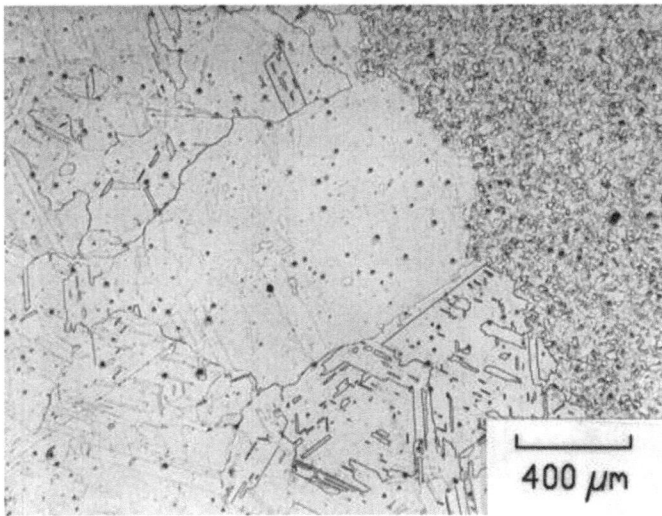

Fig. 3.6 Transverse cross-section optical micrograph of hot-rolled, induction coil solution treated and air-cooled Monel alloy K-500. Sample HR3A. Etch: NaCN, $(NH_4)_2S_2O_8$ (50×)

Fig. 3.7 Transverse cross-section optical macrograph of hot-rolled, induction coil solution treated and oil quenched Monel alloy K-500. Sample HR5O. Etch: NaCN, $(NH_4)_2S_2O_8$ $(4.25\times)$

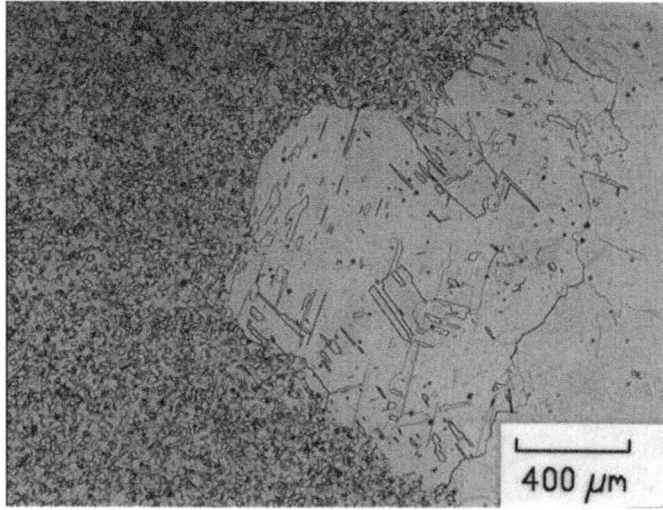

Fig. 3.8 Transverse cross-section optical micrograph of hot-rolled, induction coil solution treated and oil quenched Monel alloy K-500. Sample HR5O. Etch: NaCN, $(NH_4)_2S_2O_8$ $(50\times)$

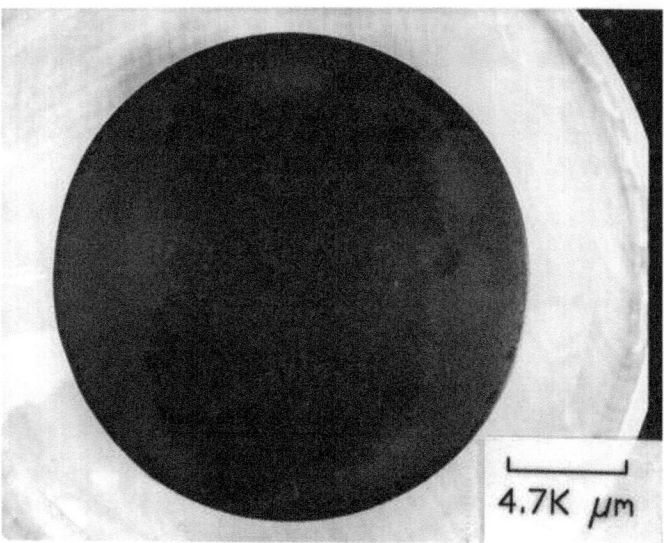

Fig. 3.9 Transverse cross-section optical macrograph of hot rolled, induction coil solution treated and water quenched Monel alloy K-500. Sample HR7W. Etch: NaCN, $(NH_4)_2S_2O_8$ (4.25×)

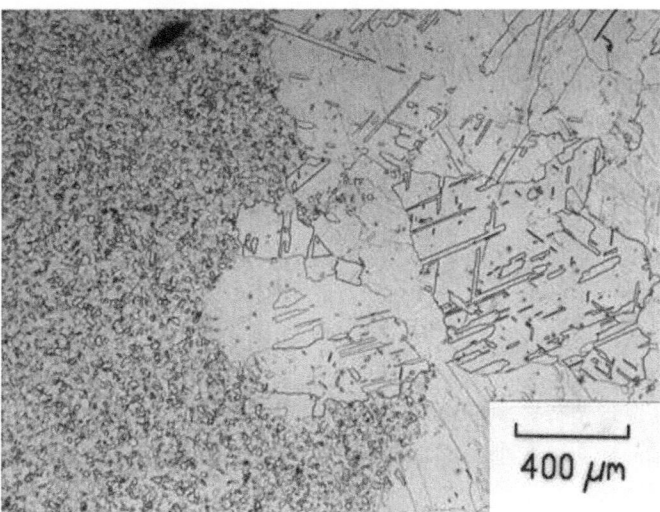

Fig. 3.10 Transverse cross-section optical micrograph of hot-rolled, induction coil solution treated and water quenched Monel alloy K-500. Sample HR7W. Etch: NaCN, $(NH_4)_2S_2O_8$ (50×)

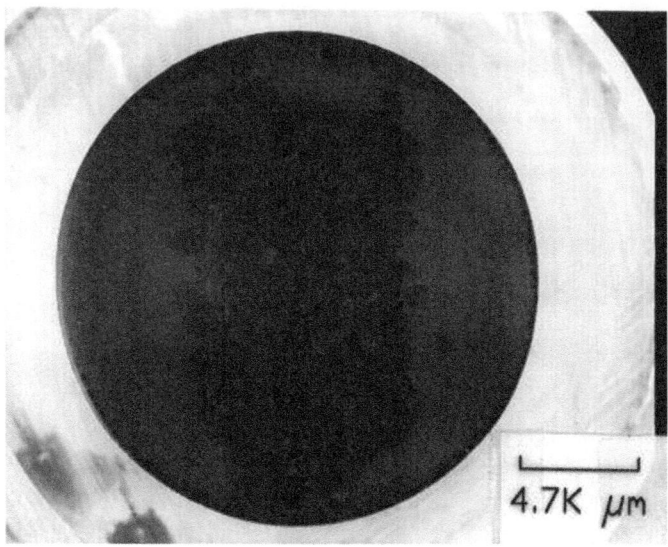

Fig. 3.11 Transverse cross-section optical macrograph of hot-rolled, induction coil solution treated and ice-water quenched Monel alloy K-500. Sample HR9I. Etch: NaCN, $(NH_4)_2S_2O_8$ (4.25×)

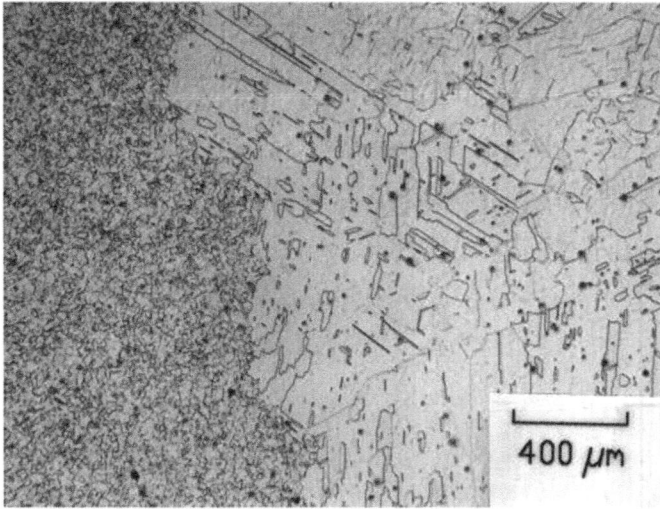

Fig. 3.12 Transverse cross-section optical micrograph of hot-rolled, induction coil solution treated and ice-water quenched Monel alloy K-500. Sample HR9I. Etch: NaCN, $(NH_4)_2S_2O_8$ (50×)

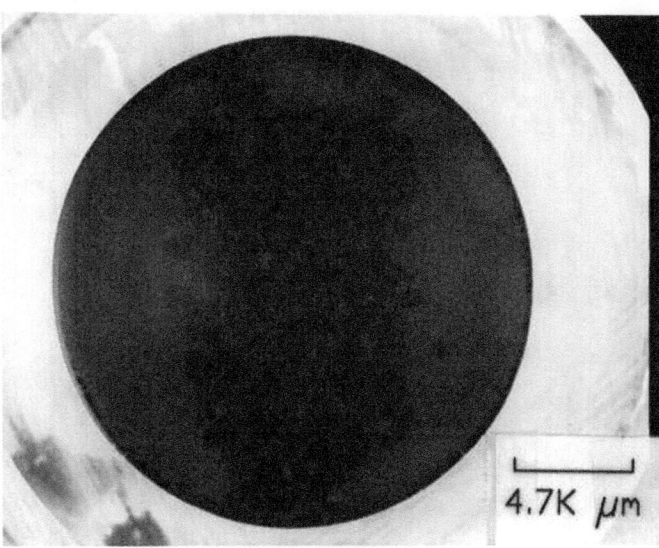

Fig. 3.13 Transverse cross-section optical macrograph of hot-rolled, induction coil solution treated and brine quenched Monel alloy K-500. Sample HR11B. Etch: NaCN, $(NH_4)_2S_2O_8$ (4.25×)

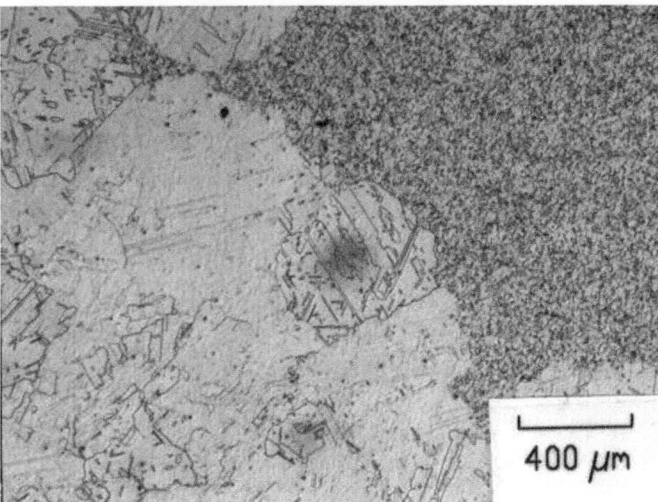

Fig. 3.14 Transverse cross-section optical micrograph of hot-rolled, induction coil solution treated and brine quenched Monel alloy K-500. Sample HR11B. Etch: NaCN, $(NH_4)_2S_2O_8$ (50×)

Fig. 3.15 Transverse cross-section optical macrograph of hot-rolled, induction coil solution treated and brine quenched Monel alloy K-500. Sample HR11B. Etch: NaCN, $(NH_4)_2S_2O_8$ ($25\times$)

Fig. 3.16 Transverse cross-section optical micrograph hot rolled, induction coil solution treated and brine quenched Monel alloy K-500. Sample HR11B. Etch: NaCN, $(NH_4)_2S_2O_8$ ($100\times$)

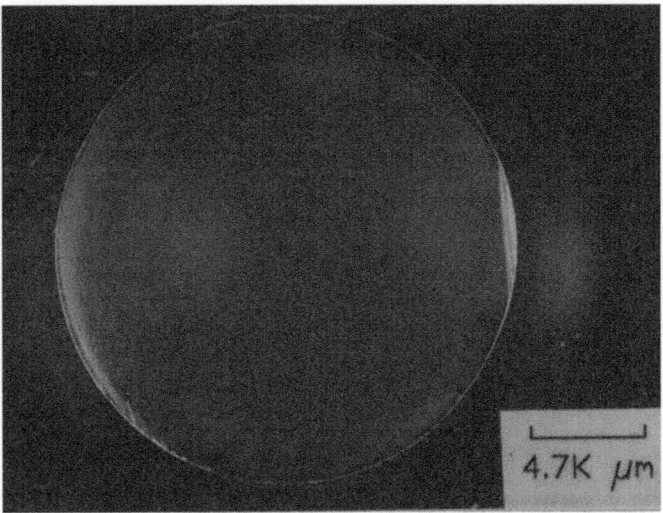

Fig. 3.17 Transverse cross-section optical macrograph of hot-rolled, air furnace solution treated and water quenched Monel alloy K-500. Sample HRWF. Etch: NaCN, $(NH_4)_2S_2O_8$ (4.25×)

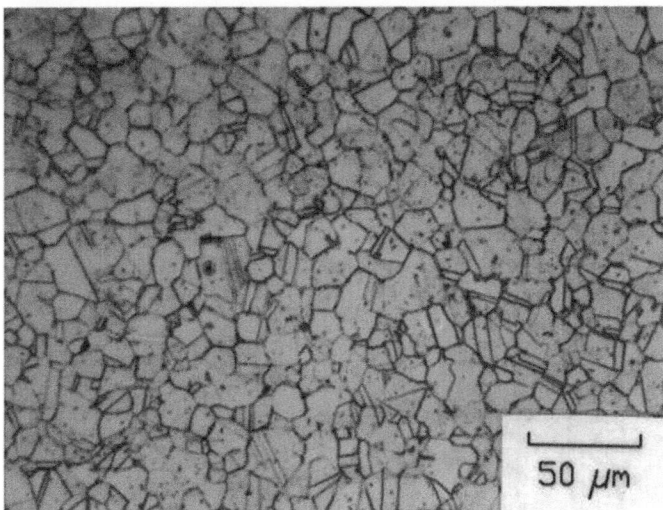

Fig. 3.18 Transverse cross-section optical micrograph of hot-rolled, air furnace solution treated and water quenched Monel alloy K-500. Sample HRWF. Etch: NaCN, $(NH_4)_2S_2O_8$ (400×)

twins and nitrides, whereas the $(NH_4)_2S_2O_8$ electrolytic etch clearly defines the grain boundaries only and omits the annealing twins.

The as-received 60,000× micrograph and vacuum furnace solution treated and furnace cooled micrograph in Fig. 3.26 lacks gamma-prime precipitates. However,

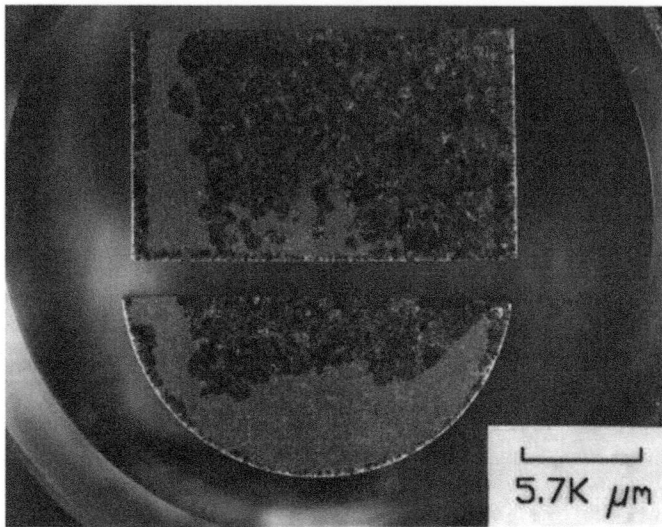

Fig. 3.19 Optical macrograph of longitudinal (upper) and transverse (lower) cross-sections of hot-rolled, induction coil solution treated and ice-water quenched Monel alloy K-500. Sample HR9I. Etch: NaCN, $(NH_4)_2S_2O_8$ (3.5×)

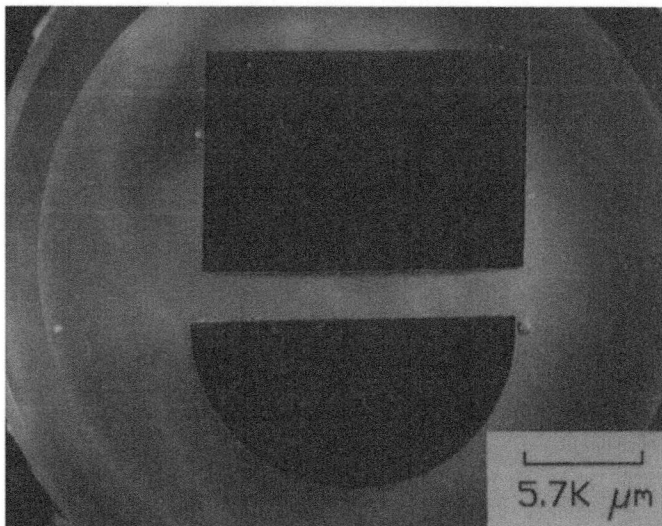

Fig. 3.20 Optical macrograph of longitudinal (upper) and transverse (lower) cross-sections of cold-drawn, induction coil solution treated, water quenched and aged Monel alloy K500. Sample CD8WA. Etch: NaCN, $(NH_4)_2S_2O_8$ (3.5×)

Fig. 3.21 Transverse cross-section optical macrograph of cold-drawn as-received Monel alloy K-500. Sample CD0. Etch: NaCN, $(NH_4)_2S_2O_8$ (4.0×)

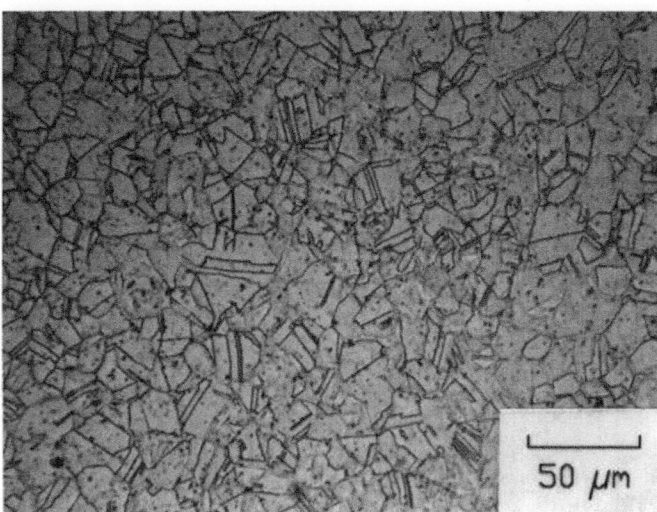

Fig. 3.22 Transverse cross-section optical micrograph of cold-drawn as-received Monel alloy K-500. Sample CD0. Etch: NaCN, $(NH_4)_2S_2O_8$ (400×)

Fig. 3.23 Longitudinal cross-section optical micrograph of cold-drawn as-received Monel alloy K-500. Sample CD0. Etch: NaCN, $(NH_4)_2S_2O_8$ (400×)

Fig. 3.24 Longitudinal cross-section optical micrograph of cold-drawn, induction coil solution treated and brine quenched Monel alloy K-500. Sample CD11B. Etch: NaCN, $(NH_4)_2S_2O_8$ (400×)

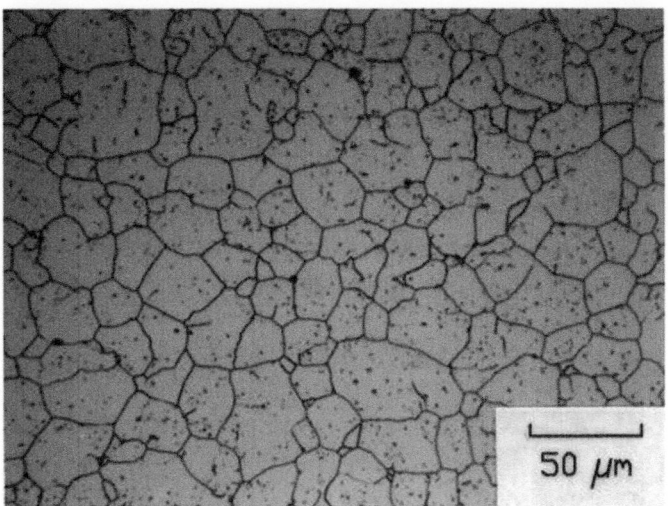

Fig. 3.25 Longitudinal cross-section optical micrograph of cold-drawn, induction coil solution treated, brine quenched and aged Monel K-500. Sample CD12BA. Etch: electrolytic $(NH_4)_2S_2O_8$ (400×)

the solution treated and aged in Fig. 3.27 micrograph clearly illustrates the spherical precipitates homogeneously distributed throughout the base matrix. The quench rate does not appear to affect neither the size nor the distribution of the gamma-prime precipitate in Monel alloy K-500 and can also be seen in Fig. 3.28 (brine quenched and aged). The needle-like features at the grain boundaries of the age hardened samples are also precipitates, although they are larger in slow cooling (Fig. 3.27) compared to rapid cooling (Fig. 3.28). There can be no other explanation for them since they are only present in age hardened materials. These precipitates are probably a variation of $Ni_3(Al, Ti)$ phase, as they are not spherical gamma-prime nor believed to be the $M_{23}C_6$-type phase precipitate described previously by Dey et al. [22]. This precipitated grain boundary phase is not considered to be a $Cr_{23}C_6$-type phase since chromium is non-existent in this alloy, as confirmed by energy dispersive X-ray spectroscopy. However, they might be TiC particles when titanium is between (0.35–0.85 %) these particles form. The composition of Ti in this alloy is 0.46 wt %, Table 2.1. TiC particles possess the FCC crystal structure and randomly precipitates along grain boundaries or within the grains [23]. The cold-drawn material was observed to have a uniform microstructure irrelevant of being solution treated in either a vacuum furnace or an induction coil, except for a few isolated islands of abnormally large or coarse grains (AGG).

Annealing twins that usually appear as parallel sided bands, were abundant in the solution treated samples. These microstructure features are known to be prominent in recrystallized face-centered-cubic (FCC) metals and alloys where the interfacial energies associated with coherent twin boundaries are extremely smaller than those associated with the grain boundaries [24]. An annealing twin is formed in a crystal

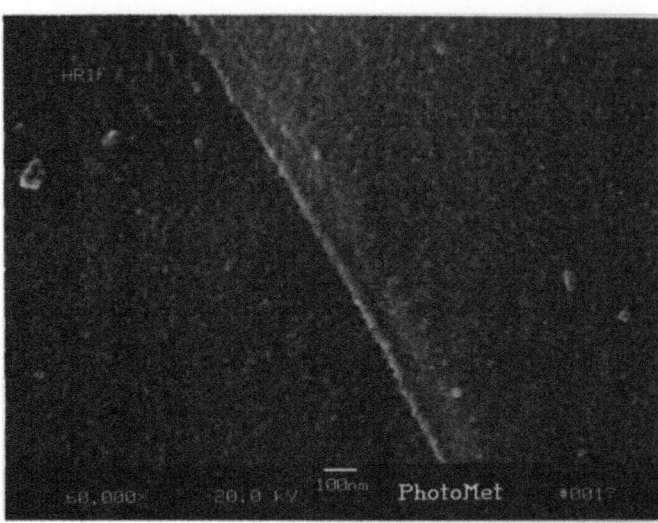

Fig. 3.26 Transverse cross-section SEM micrograph of hot-rolled, vacuum furnace solution treated and furnace cooled Monel alloy K-500. Sample HR1F. Etch: NaCN, $(NH_4)_2S_2O_8$ (60,000×)

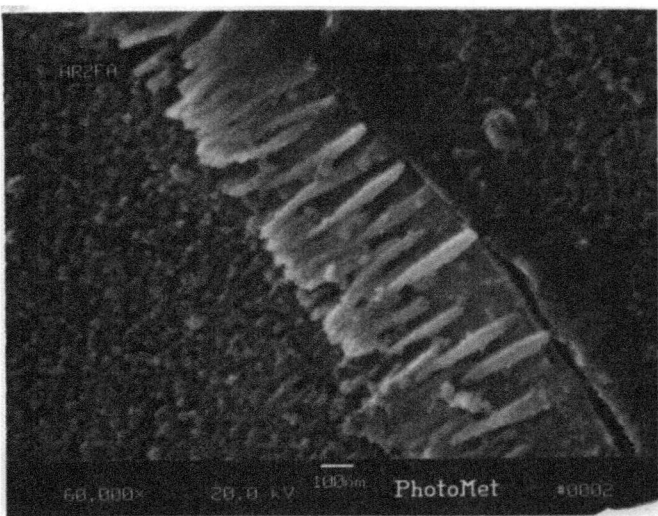

Fig. 3.27 Transverse cross-section SEM micrograph of hot-rolled, vacuum furnace solution treated, furnace cooled and aged Monel alloy K-500. Sample HR2FA. Etch: NaCN, $(NH_4)_2S_2O_8$ (60,000×)

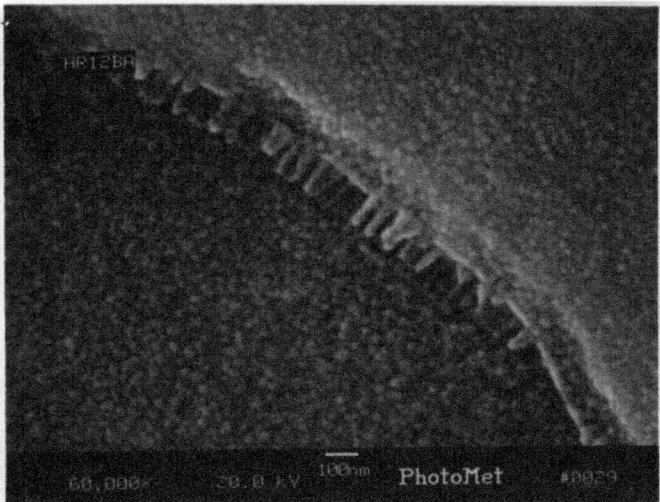

Fig. 3.28 Transverse cross-section SEM micrograph of hot-rolled, induction coil solution treated, brine quenched and aged Monel alloy K-500. Sample HR12BA. Etch: NaCN, $(NH_4)_2S_2O_8$ $(60,000\times)$

during recrystallization and is defined as a feature where two portions of a crystal have a definite crystallographic relationship.

3.4 Induction Heat Treating

Induction heat treating is accomplished by passing an alternating current through a work coil. As a result of this current, a highly concentrated and rapidly alternating magnetic field is generated within the coil. The magnetic field couples with the part within the coil and induces an electrical potential within the part. Under this condition, since the part itself represents a closed circuit, the induced voltage causes current flow. As the part resists the flow of the induced current, the part heats up [25]. For induction heat treating, the depth of current penetration into the part is affected by three variables: material permeability, resistivity, and the greatest, the alternating current frequency. The significance of this is that the 450 kHz frequency used in this study is at the high end of frequencies used for this activity, the reason being that the depth of current penetration decreases as frequency increases. As a result, the induced currents in the part resulted in resistive heating of the outermost periphery of the part and then the rapid conduction of heat to the core of the sample. For solution treating, it is not clear if the additional 3–7 min that it took for the core of the sample to reach the solution treatment temperature would have a significant impact. For the hot-rolled material, the impact of this heating method was most pronounced with the bimodal grain structure (AGG).

3.5   Miscellaneous Particles

Titanium nitride/carbide particles were abundantly and randomly distributed throughout the matrix of Monel alloy K-500. These particles, typically cubicle in shape, are visible at relatively low optical and SEM magnifications. The effect of these particles has generally been studied to have a negligible impact on the properties of the alloy [26].

3.5.1   Precipitates

The gamma-prime phase (Ni_3 (Al, Ti)) precipitate is exceptionally well suited in Monel alloy K-500, due to its FCC crystal structure and lattice constant (approximately 0.1% mismatch) with nickel that permits homogeneous nucleation of a precipitate with low surface energy and long duration stability. This small lattice mismatch is responsible for the spherical precipitates. The gamma-prime precipitate occurs as spheres between 0 and 0.2% mismatch, cubes at approximately 0.5–1.0% mismatch, and then plates above approximately 1.25%. Figures 3.28 and 3.29 support the mismatch level for Monel alloy K500, as the precipitates are spherical. The coherency between the gamma matrix and gamma-prime precipitate is maintained by tetragonal distortion [27].

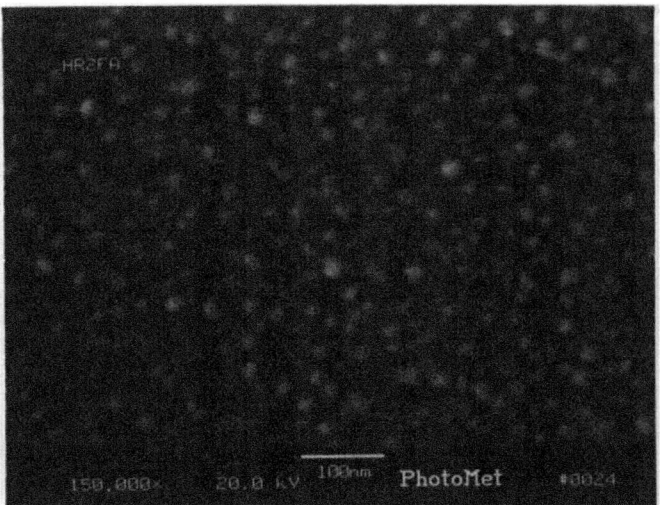

Fig. 3.29 Transverse cross-section SEM micrograph of hot-rolled, vacuum furnace solution treated, furnace cooled and aged Monel alloy K-500. Sample HR2FA. Etch: NaCN, $(NH_4)_2S_2O_8$ (150,000×)

Typical transmission electron microscopy (TEM) micrographs are illustrated in Figs. 3.30, 3.31 and 3.32. As with the SEM micrographs, the gamma-prime precipitates are clearly not present in the as-received materials or solution treated only samples (Fig. 3.30), while they are clear in the age hardened samples (Figs. 3.31 and 3.32). Again, as with the high magnification SEM micrographs, Fig. 3.32 illustrates the second precipitating phase at the grain boundary, in addition to the gamma-prime in the base matrix.

Fig. 3.30 Transverse cross-section TEM micrograph of hot-rolled as-received Monel alloy K-500. Sample HR0 (8000×)

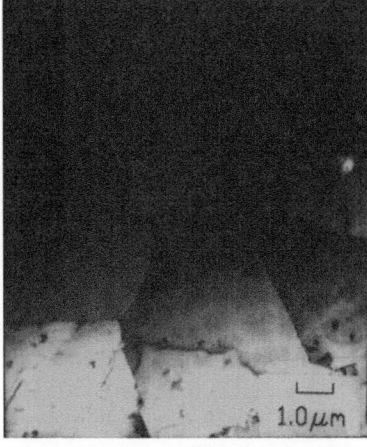

Fig. 3.31 Transverse cross-section TEM micrograph of cold-drawn, induction coil solution treated, air cooled and aged Monel alloy K-500. Sample CD4AA (48,000×)

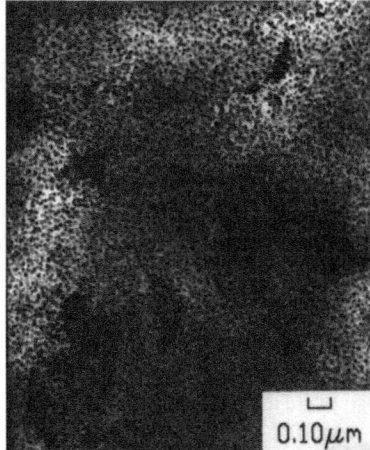

Fig. 3.32 Transverse
cross-section TEM
micrograph of hot-rolled,
induction coil solution
treated, water quenched and
aged Monel alloy K-500.
Sample HR8WA (60,000×)

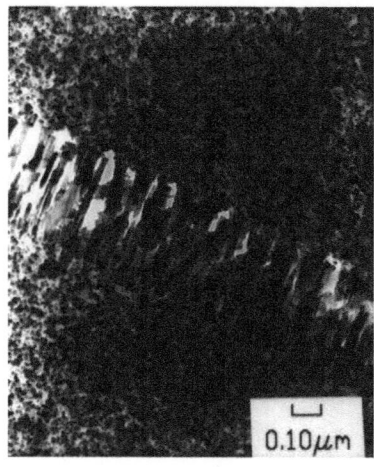

3.6 Crystallographic Texture

The as-received and few processed samples of nickel Monel alloy K-500 were tested
and analyzed using the Schultz reflection method of X-ray diffraction on a Scintag's
X1 advanced Diffraction System. Scintag DMSNT 1.35a software and the Preferred
Orientation Package—Los Alamos (POPLA) [19], were used in the texture analysis
of the samples. This analysis produced pole figures, inverse pole figures, recalculated
pole figures, and orientation distribution functions (ODF's) for each material. These
ODF's were used to determine the deformation, shear, and recrystallization components of each sample's texture. Six samples were selected, these are the as-received
cold drawn (CD0), cold drawn, solution treated, and brine quenched (CD11B) and
cold drawn, solution treated, brine quenched and age hardened CD12BA. The hot
rolled as- received and treated in the same way were also selected, HR0, HR11B,
HR12BA, seen in Figs. 2.2 and 3.33.

The deformation and shear and the recrystallization components extracted from
the ODF's for the hot rolled samples HR0, HR11B and HR12BA are plotted in
Fig. 3.34a and b. The deformation and shear components are the brass {110} <112>,
copper {112} <111>, shear {111} <112>, x {110} <111>, and rotated brass {110}
<223>. The recrystallization components are the P {110} <122>, RCT {013} <013>,
RCN {001} <013>, cube {001} <100> and Goss {110} <001>.

From Fig. 3.34a and b, there is an increase in the copper deformation component {112} <111>, the rotated brass deformation component {110} <223> and the
cube recrystallization component {001} <100> from the as-received hot rolled material (HR0) to the solution treated brine quenched (HR11B) to the age hardened
(HR12BA). The deformation and shear components, as well as the recrystallization
components of all six samples that were cold-rolled (CD0, CD11B and CD12BA)
and hot-rolled (HR0, HR11B and HR12BA), are plotted in Fig. 3.35a and b. However,

it is clear from Fig. 3.35a and b that all these components are weak with 6× or 7× the random texture.

In Orientation Imaging Microscopy (OIM), electron backscattered patterns (EBSP), formed in a scanning electron microscope, are collected from points on the sample surface distributed over a regular grid and automatically indexed. From this data, a map was constructed displaying changes in crystal orientation over the specimen surface. The orientation of each point in the microstructure is known and hence the location, line length and type of boundaries. An ElectronScan ESEM was used for the EBSP experiments, and OIM software (from TSL, Provo Utah) was used to index the EBSPs and to produce microstructure maps. The results showed two distinct microstructures. The cold drawn microstructure showed a reasonably textured small sized grain microstructure whereas the hot rolled sample showed a recrystallized large grained microstructure.

3.7 Discussion

AGG is a mode of grain growth which was investigated for more than 70 years [10]. Yet understanding the mechanisms is far from being complete. Probably the Fe-3% Si is the most studied alloy for its AGG with a strong Goss orientation {110} <001> texture evolving. Arguments emerged about why the boundaries of Goss grains travel faster than matrix grains overcoming pinning force by precipitates or if some grains have non-Goss orientation would grow abnormally if they are made to have sub-boundaries by any means. These and other underlying mechanisms have been studied by different scientists from 1935 [10] until the present [28–31].

Nickel based alloys deserve closer attention. Nickel alloys are found in aircraft engines, land-based power turbines and nuclear and chemical processing plants. They contain several precipitates of varying sizes and compositions. It is essential to understand the component properties, failure modes and identify long-term performance concerns such as AGG which can be detrimental to mechanical properties [32–36]. AGG for example caused the fatigue life of nickel-based super alloy 718 to decrease an order of magnitude [33]. In another example, K Monel 500 (Ni-Cu-Al alloy), used in this study, was the subject of arbitration between companies [15]. Bolts made of K Monel 500 passed the random testing and others did not. The problem was attributed to improper heat treatment which caused excessive grain growth in the bolts that failed testing. The problem is even in the bolts that passed testing a bimodal grain structure with AGG was evident.

Dennis et al [13] and Miller et al [36] listed the possible root causes or factors linked to cases of AGG. These are the presence of large pre-existing grains, dissolution or coarsening of precipitates, solute drag, boundary anisotropy and faceting, effect of crystallographic texture and strain. A summary of each and how it relates to K Monel 5000 in this study will be presented.

The presence of large pre-existing grains in the hot rolled as-received samples, Figs. 3.1 and 3.2 (14 μm in grain size, Table 3.3) and in the cold drawn as-received

Fig. 3.33 Samples selected
for texture measurements

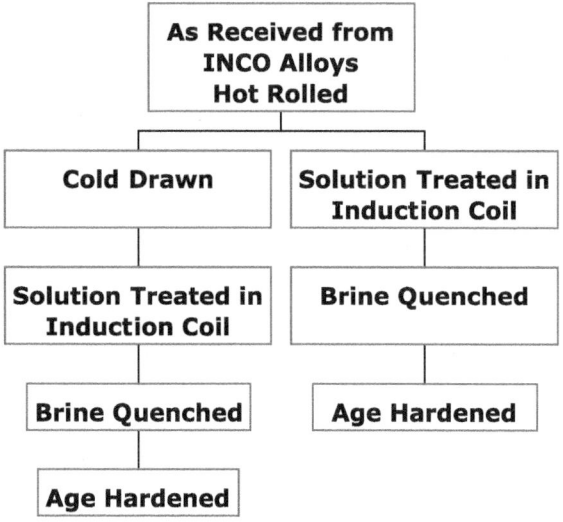

samples Figs. 3.21 and 3.22 (21 μm in grain size) are not detected. The initial grain size in both conditions is fine (14–20 μm) and uniform. For the dissolution or coarsening of precipitates leading to localized reduction in pinning boundaries [13, 36], there was no evidence of this, Figs. 3.26, 3.27, 3.28, 3.29, 3.30, 3.31 and 3.32. Solute drag (Zener effect) [37], boundary anisotropy and faceting were not observed. For crystallographic texture, the deformation/shear components, and the recrystallization components in the hot rolled and cold drawn conditions were slightly different. However, the texture components were very weak (5–7× as random) and could not explain AGG in the hot rolled condition and the absence of it in the cold drawn condition, Figs. 3.33, 3.34 and 3.35.

The strain effect, particularly effect of small deformations was reported in several studies [32, 35, 36, 38–42]. It is the effect that explains the occurrence of the abnormal or coarse grains in the hot rolled material in this study. Underwood et al [32] reported that AGG occurred at specific pairing of strain and temperature and provided an empirical relation describing the effect of thermal exposure and strain content on the onset of AGG in Ni200. Agnoli et al [35] indicated that when the initial micro-structure of Inconel 718 was strained ($\varepsilon < 0.1$) before annealing then low stored energy grains grow to a large extent. They called the phenomena strain-induced selective grain growth in a pinned microstructure which results in a bimodal grain size distribution. The appearance of the coarse grains should be referred to as a primary recrystallization phenomenon resulting from the presence of very few deformed grains in the microstructure before annealing. Miller et al. [36], appears to agree with Angoli et al. [35], noted that the largest grains observed were ~700 μm in diameter after 60 s of heat treatment at a deformation of around 2%. Above this threshold the recrystallized grains decrease, and recrystallization nuclei decrease with an increase in deformation percent. The inhomogeneity of deformation at a

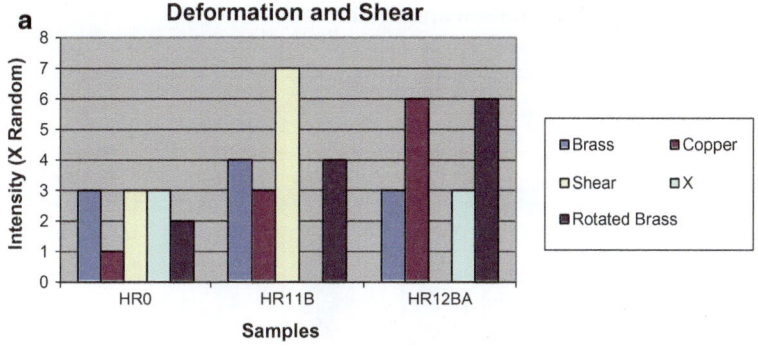

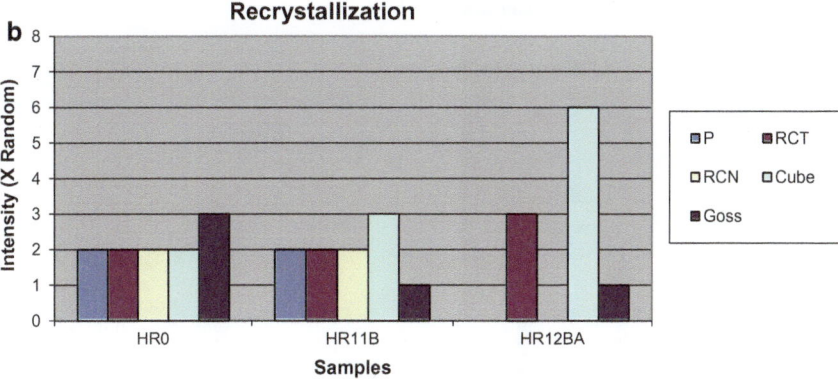

Fig. 3.34 a Deformation and shear components. **b** Recrystallization components

very low strain creates few nuclei and this promotes a very coarse recrystallize grain. Miller et al. [36] proposes that abnormally large grains in Rene 88 material is probably due to nucleation limited recrystallization and not abnormal grain growth.

Sarrail et al. [40] and Rodriguez et al. [41] used the Sauvert hypothesis proposed in 1935 to analyze abnormal grain growth in Zr 702 alloy. Sauvert [39] indicated that abnormal grain growth occurs at certain strains, and critical strains. Above or below these, abnormal grain growth does not occur. Several nuclei, of which he called germes, will give rise to grains. In the under-strained or overstrained conditions, abnormal grain growth will not occur. At the critical strain condition these nuclei will give rise upon annealing to a few large grains. In the overstrained condition a great number of these nuclei will produce, upon annealing, many fine grains. The work done on Zr 702 alloy was on bent samples [40, 41] where there was a non-uniform strain from maximum compression to the neutral axis and then to maximum tension. In hot-rolling only, a small variation of strain can occur within the sample.

The 30% additional cold work in the initial cold drawn samples in this study may induce enough strain for primary recrystallization to occur after the solution treatment without resulting coarse grains. In primary recrystallization, new grains form, while

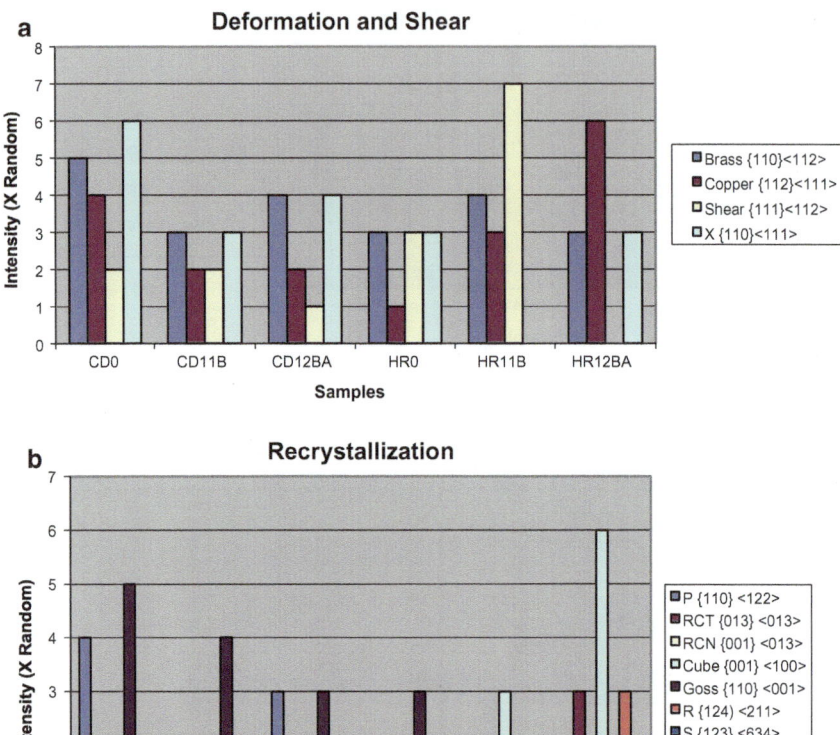

Fig. 3.35 a Deformation and shear texture components of all six samples. **b** Recrystallization texture components of all six samples

in strain induced grain growth in the hot rolled material new grains are not created and the strains are insufficient to promote new grains after the solution treatment. The combined effects of the presence of lower strain of the hot rolled samples coupled with a very rapid rate of heating (450KZ induction), appears to have promoted the abnormal grains. These grains are most likely formed due to AGG; however, they could also be coarse grains due to a primary recrystallization phenomenon resulting from the presence of very few deformed grains in the microstructure before annealing, Agnoli et al. [35] and Miller et al. [36] or due to Sauvert [39] critical strain theory.

Chapter 4
Conclusions

1. Monel alloy K-500's mechanical properties, based on hardness testing, are not affected by quench rates ranging from 0.6 °C/s (1 °F/s) as obtained in a furnace cool to 171 °C/s (308 °F/s) as obtained in a brine quench from the solution treat temperature of 1010 °C (1850 °F). Further, the initial material conditions of either hot-rolled or cold-drawn do not affect the final properties of the material.
2. There is no distinguishable difference in the size and distribution of second phase hardening constituents (gamma-prime) due to quench rate. This visual assessment of the precipitates is substantiated by equivalent hardness measurements.
3. Hot-Rolled Monel alloy K-500 does not solution treat properly by high frequency (450 kHz) induction heat training. This heat-treating method results in a severe duplex microstructure with grain sizes varying by orders of magnitude. The very large or coarse grains present in the microstructure are referred to by industry personnel as "germinated" or "Elephant" grains.
4. Solution treating hot-rolled Monel alloy K-500 in an air furnace avoids a nonuniform microstructure.
5. Solution treating cold-drawn or hot-rolled Monel alloy K-500 in a vacuum furnace (radiative heat transfer) result in a few random islands of coarse grains.
6. Solution treating cold-drawn Monel alloy K-500 by high frequency (450 kHz) induction eat treating (resistive heating) results in a few random islands of coarse grains.
7. The cause of the coarse grains is attributed to the combination of the initial starting condition of the hot rolled material (low strain) with the high frequency induction solution treatment (rapid heating rate).
8. The reason for the development of the coarse grains can be AGG or very coarse recrystallized grains or due to the Sauvert critical strain theory.
9. There is no distinguishable difference in the precipitate size and distribution in the base matrix or at the grain boundaries whether the grain be fine or coarse.

© The Author(s), under exclusive license to Springer Nature Switzerland AG 2023
R. J. Reidel et al., *On the Abnormal/Coarse Grain Formation in K-Monel 500 Alloy*,
SpringerBriefs in Continuum Mechanics,
https://doi.org/10.1007/978-3-031-31079-9_4

Also, there is no evidence of precipitate depletion of the base matrix near nitride particles.

10. The deformation/shear and recrystallization components of the hot-rolled and cold drawn conditions are slightly different but too weak to be responsible for the different responses to the solution treatment by the induction furnace.

References

1. Shirdel M, Mirzadeh H, Parsa MH (2014) Abnormal grain growth in AISI 304L stainless steel. Mater Charact 97:11–17
2. MacPherson RD, Srolovitz DJ (2007) The von Neumann relation generalized to coarsening of the three-dimensional microstructures. Nature 446:1053–1055
3. Koo J, Yoon D (2001) Abnormal grain growth in bulk Cu—the dependence on initial grain size and annealing temperature. Metall Mater Trans A 32:1911–1926
4. Choi JS, Noon DY (2001) The temperature dependence of abnormal grain growth and grain boundary faceting in 316 stainless steel. ISIJ Int 41:478–483
5. Humphreys J, Rohrer GS, Rollett A (2017) Grain growth following recrystallization. In: Recrystallization and related annealing phenomena, chapter 11. pp 375–429
6. Blendwell JE, Handwerker CA (2001) Ceramics: grain growth. In: Encyclopedia of materials: science and technology, 2nd edn. pp 1105–1108
7. Guo Y, Liu Z, Huang Q, Lin CT, Du S (2018) Abnormal grain growth of UO2 with pores in the final stage of sintering: a phase field study. Comput Mater Sci 145:24–34
8. Martin JW (2001) Grain growth, abnormal. In: Encyclopedia of materials: science and technology, 2nd edn. pp 3634–3636
9. Bauri R, Yadav D (2018) Processing nonequilibrium composite (NMMC) by FSP. In: Metal matrix composites by friction stir processing. Butterworth Heinemann
10. Goss N (1935) New development in electrical strip steel characterized by fine grain structure approaching the properties of a single crystal. Trans Am Soc Metals 23:511–531
11. Homma H, Hutchinson B (2003) Orientation dependence of secondary recrystallization in silicon-iron. Acta Mater 51:3795–3805
12. Fang F, Zheng YX, Lu X, Wang Y, Lan MF, Yuan G, Misra RDK, Wang GD (2018) Abnormal growth of 100 grains and strong cube texture in strip cast Fe-Si electrical steel. Scripta Mater 147:33–36
13. Dennis J, Bate PS, Humphreys FJ (2009) Abnormal grain growth in Al-3.5 Cu. Acta Mater 57:4539–4547
14. Jung S-H, Yoon DY, Kang S-JL (2013) Mechanisms of abnormal grain growth in ultrafine-grained nickel. Acta Mater 6:5685–5693
15. Zakharia K, Foyos J, Marloth R, Es-Said OS (2000) Failure analysis of K-Monel 500 (Ni–Cu–Al alloy) bolts. Eng Fail Anal 7:323–332
16. Federal Specification QQ-N-00286F (SH) (1990) Nickel-copper-aluminum alloy. Wrought, pp 6
17. High Performance Alloys, I.-S.P (n.d.) Monel K500 (UNS N05500). https://www.hpalloy.com/Alloys/descriptions/MONELK_500.aspx
18. Halle A (1971) Grain size measurement by the intercept method. Metallography 59–78

© The Editor(s) (if applicable) and The Author(s), under exclusive license to Springer Nature Switzerland AG 2023
R. J. Reidel et al., *On the Abnormal/Coarse Grain Formation in K-Monel 500 Alloy*, SpringerBriefs in Continuum Mechanics, https://doi.org/10.1007/978-3-031-31079-9

19. Kallend J, Kocks UF, Rollett AD, Wenk HR (1991) PopLA—an integrated software system for texture analysis. Texture and Microstruct 14–18:1203–1208
20. Tillack DJ, Manning JM, Hensley JR (1991) ASM handbook, volume 4, heat treating, heat treating of nickel and nickel alloys. pp 910–911
21. Reed-Hill RE (1973) Physical metallurgy principals, chapters 7–9. pp 267–375
22. Dey GK, Tewari R, Rao P, Wadekar SL, Mukhopadhyay P (1993) Precipitation hardening in nickel-copper base alloy Monel K 500. Met Trans A 24A:27092719
23. Marenych O, Kostryzheu A (2020) Strengthening mechanisms in nickel-copper alloys: a review. Metals 10:1358. https://doi.org/10.3390/met1010358
24. Dey GK, Mukhopadhyay (1986) Precipitation in the NI–Cu-base alloy Monel K-500. Mater Sci Eng 84:177–189
25. Stevens N (1981) Metals handbook, 9th edn, volume 4, heat treating, p 451
26. Farag MM, Hamdy MM (1976) Behavior of a nickel-base high-temperature alloy under hot-working conditions. Met Trans 7A:221–228
27. Stout JJ, Crimp MA (1992) Abnormal grain growth in textured Fe–Al intermetallics. Mater Sci Eng, A 152:335–340
28. Braun C, Dake J, Kril CE, Birringer R (2018) Abnormal grain growth mediated by fractal boundary migration at the nanoscale. Sci Rep 8:1592
29. Kim T-Y, Shim HS, Choi S, Na T-W, Kwon D, Hwang N-M (2019) Synchrotron X-ray microdiffraction analysis of abnormally growing grains induced by indentation in Fe-3% Si steel. Mater Charact 156:109845
30. Kim T-Y, Kim H-K, Jeong Y-K, Ahn Y-K, Shim H-S, Kwon D, Hwang N-M (2020) Ex-situ sequential observation on inland and peninsular grains in abnormally growing Goss grains in Fe-3% Si Steel. Metal Mater Int
31. Kim T-Y, Na T-W, Shim H-A, Ahn Y-K, Jeong Y-K, Han HN, Hwang N-M (2020) Misorientation characteristics at the growth front of abnormally-growing gas grains in Fe–Si steel. Metal Mater Int
32. Underwood O, Madison J, Thompson G, Welsh S, Evans J An examination of abnormal grain growth in low strain Ni-200, metallography microstructure. Analysis 5(206):302–312
33. Flageolet B, Yousfi O, Dahan Y, Villechaise P, Cormier J (2010) Characterization of microstructures containing abnormal grain growth zones in alloy 718. In: Ott E, Groh J, Banik A, Dempster I, Gabb T, Helmink R, Liu X, Mitchell A, Sjoberg GP, Wadatowska-Sarnek A (eds) Conference: 7th international symposium on superalloy 718 and derivations At Pittsburgh, PA, TMS, p 595–606
34. Noell P, Worthington D, Taleff E (2017) Growth of pre-existing abnormal grains in molybdenum under static and dynamic conditions. Mat Sci Eng A 692:2434
35. Agnoli A, Bernacki M, Loge R, Franchet J, Laigo J (2015) Selective grow of low stored energy grains During δ sub-solvus annealing in the Inconel 718 nickel-based superalloys. Met Mat Trans A 46A:4405–4421
36. Miller V, Johnson A, Toret C, Pollock T (2016) Recrystallization and the development of abnormally large grains after small strain deformation in a polycrystalline nickel-based superalloy. Metal Mater Trans A 47A:1566–1574
37. Hibbard G, Radmilovic V, Aust K, Erb U (2008) Grain boundary migration during abnormal grain growth in nanocrystalline Ni. Mat Sci Eng A 494(1–2):232–238
38. Bozzolo N, Agnoli A, Souai N, Bernacki M, Loge R (2013) Strain induced abnormal grain growth in nickel-based superalloys. Mat Sci Forum 753:321–324
39. Sauvert A (1935) The metallography and heat treatment of iron and steel. Chapter XIX, pp 285–292

40. Sarrail B, Schrapps C, Babakhanyan S, Muscare K, Foyos J, Ogren J, Sparkowich S, Sutherlin R, Hilty J, Clarkland R, Es-Said OS (2007) Annealing and anomalous (biomedical) grain growth of Zr 702. Eng Fail Anal 14:652
41. Rodriguez N, Dickinson T, Huy Nguyen D, Park E, Foyos J, Sutherlin R, Sparkowich S, Hogue F, Stoyanov P, Ogren J, Plotkin E, Es-Said OS (2008) On the bimodal grain growth in zirconium grade 702 alloy. Eng Fail Anal 15:440
42. Randle V, Brown A (1988) The effect of strain on grain misorientation texture during grain growth incubation period. Philos Mag A 58:717

Printed by Printforce, the Netherlands